KOALA PUBLICATIONS

The Solar System For Kids 6-8 Adventure

The Young Astronomer's Starter Guide To Learning About the Universe

Copyright © 2023 by Koala Publications

All rights reserved. No part of this publication may be reproduced, stored or transmitted in any form or by any means, electronic, mechanical, photocopying, recording, scanning, or otherwise without written permission from the publisher. It is illegal to copy this book, post it to a website, or distribute it by any other means without permission.

Koala Publications asserts the moral right to be identified as the author of this work.

Koala Publications has no responsibility for the persistence or accuracy of URLs for external or third-party Internet Websites referred to in this publication and does not guarantee that any content on such Websites is, or will remain, accurate or appropriate.

Designations used by companies to distinguish their products are often claimed as trademarks. All brand names and product names used in this book and on its cover are trade names, service marks, trademarks and registered trademarks of their respective owners. The publishers and the book are not associated with any product or vendor mentioned in this book. None of the companies referenced within the book have endorsed the book.

First edition

This book was professionally typeset on Reedsy. Find out more at reedsy.com

Contents

1	Introduction	1
2	Exploring the Universe	2
3	Meet the Sun	12
4	Meet the Moon	15
5	The Eclipse	21
6	The Solar System	26
7	The Inner & Outer Planets	33
8	Exploring The Planets	36
9	Pluto & The Kuiper Belt	46
10	The Shooting Star	50
11	Conclusion	53
12	References	54

1

Introduction

Hello there, space explorers! Are you ready for a super fun adventure to learn all about our Solar system?

With this book, you'll get to blast off and discover all sorts of cool stuff about the planets that are way up in the sky! You'll get to ride in a rocket ship and fly past each planet to learn all about their unique features and personalities. Did you know that Jupiter is the biggest planet in our Solar System or that Mercury is the closest planet to the Sun? You'll also get to see how all the planets move and dance around each other in space. So, let's get ready to launch! Put on your space suit and helmet, and buckle up tight for an amazing journey through our cosmic neighborhood. Let's explore the wonders of the universe together!

2

Exploring the Universe

Let's get started on this exciting adventure of exploring the universe!

Are you ready to learn about all the amazing things that exist in this huge, vast, and mysterious place? Get ready to use your imagination and let's take a journey together into the wonders of the universe!

The universe is a huge place that's full of all kinds of things! It's like a giant room that goes on and on forever. Imagine you're standing in the middle of this room, and you can look in every direction and see something new and exciting.

One of the things you might see is bright, shiny objects that twinkle in the sky. These objects are very far away, but they look like they're close together because they're so bright. They come in all different colors and sizes.

Another thing you might see is big swirling clouds that are made up of gas and dust. These clouds are so big that they can be hundreds of thousands of times bigger than the Earth! Scientists call them nebulae, and they're places where new stars are born.

There are also huge groups of objects that travel together through space. These groups are called galaxies, and they can have billions of objects in them! Galaxies come in all different shapes and sizes, and they can be really beautiful.

Finally, there are objects that move through space by themselves, like big rocks or chunks of ice. These objects are called asteroids and comets, and they can range in size from very big to very small.

So, that's the universe! It's a huge place that's full of all kinds of amazing things. Even though we can't explore it all, it's still fun to imagine what's out there and what we might discover next!

So, the next time you gaze up at the night sky, remember that you're looking at just a small part of the universe. So keep looking up, keep dreaming and keep exploring - the universe is waiting for you!

Did you know that scientists use really cool tools to learn about the planets and other things in space? They have these big binoculars called telescopes that can see things that are really far away, like planets and stars! Scientists use telescopes to learn about the planets in our Solar System, like how big they are and what they're made of.

But sometimes, telescopes aren't enough. So, scientists built special spaceships that can travel through space and visit the planets! These spaceships have cameras, instruments, and other cool tools that can study the planets up close! That's like having a big magnifying glass that can see things in space!

Did you know that people have been exploring space for a long, long time? The first person to travel to space was a man named Yuri Gagarin, and he did it in 1961. Since then, people have landed on the Moon an even sent robots to explore Mars. It's like we're on a big adventure, exploring all the amazing things that are out there!

By studying the Solar System, scientists hope to learn more about how it formed and how it works.

Exploring space is not only super fun and exciting, but it also

The spacesuit helmet is one of the most important parts of the outfit. It's like a big bubble that covers the astronaut's head. Inside the helmet, there's air to breathe, just like you're breathing right now. But the spacesuit helmet also has a clear shield that helps protect the astronaut's face from space rocks and other things that could hurt them. It's like a superhero shield!

The suit part of the spacesuit is also super awesome. It's like a big, thick, and flexible outfit that covers the astronaut's body. It's like a spacesuit superhero costume! The suit keeps astronauts warm or cool depending on the temperature in space. And it also has a special feature that helps the astronaut move around in space. It's like the suit gives the astronaut special space powers!

So that's why astronauts need special suits and helmets when they go into space. It's like they're wearing a cool superhero outfit that keeps them safe and allows them to do their job. They're like brave explorers, venturing into the unknown and discovering amazing things about space.

I hope you had a blast learning about spacesuits and space exploration today young space enthusiasts! Keep exploring and dreaming, and who knows, maybe one day, you'll be the one wearing a spacesuit and exploring the mysteries of space!

So there you have it little astronauts! We've explored the vast wonders of the universe and learned about cool tools that scientists use to learn about the planets and other things in space. Learned about gravity and discovered why astronauts need special suits to survive in space.

Now, let's see how much you've remembered with three fun questions:

1 - What are some of the things you might see in the universe?

(10 seconds to think answer)

Answer:

3

Meet the Sun

Hello again, young astronomers! Are you excited to learn about the star that's at the center of our Solar system? It's called the Sun, and it's a really big ball of hot gas that gives us light and heat every day.

It's also really, really hot - so hot that if you tried to touch it, you would get burned up in an instant!

But even though the Sun is far away from us, we still feel its warmth and light every day. The Sun helps plants grow, and it's what makes our sky turn from blue to pink to orange during a beautiful sunset.

Did you know that the Sun is also very important for the Earth's climate and weather? It gives us energy for things like wind and waves, and it even affects the temperature of our planet.

Now that we've met the Sun let's find out what it's made of! The Sun is like a giant gas balloon that's filled with really special gases. Imagine it like a big party balloon in the sky!

The Sun is mostly made of two special gases called hydrogen and helium. These gases are like the Sun's superpower that makes it shine so bright. They get so hot and squished together in the middle of the Sun that they make a giant fire. It's like a big firework party!

The fire inside the Sun is so hot and bright that it makes light and heat that travel all the way to Earth. That's what we see as sunlight! And do you know what's even cooler? The energy from the Sun helps us make wind and waves and even keeps the Earth's temperature just right.

But the Sun isn't just made of hydrogen and helium. It's like a big treasure chest in the sky, filled with other cool gases like carbon, nitrogen, and oxygen. These elements are what make up our whole Solar system, including our home planet, Earth!

So, there you have it, space adventurers! The Sun is like a super cool, giant gas balloon in the sky that's filled with amazing elements that keep us warm and light up our world.

C'mon kids, it's question time again! Let's answer a couple of questions together shall we?

1 - What are the two special gases that make up the majority of the Sun?

(10 seconds to think and answer)

They are Hydrogen and Helium.

2 - What are some other gases that can be found in the Sun besides hydrogen and helium?

(10 seconds to think and answer)

They are Carbon, Nitrogen and oxygen.

4

Meet the Moon

Now that we've learned about the Sun let's travel a little further and discover another amazing celestial body - The Moon!

Have you ever heard of a celestial body? It's like a giant space rock but even more amazing! Let me tell you all about it!

A celestial body is anything in space that isn't on planet Earth. That means the Moon, the Sun and there are even more celestial bodies out there that we can't see with our eyes, like comets, asteroids, and planets orbiting other stars far, far away.

These celestial bodies are like big, beautiful jewels in the sky. They twinkle and shine, and some of them even have their own special names.

Scientists love studying celestial bodies because they can tell us so much about the universe we live in. By looking at the light and radiation they give off, we can learn about their size, their temperature, and even what they're made of - how amazing is that right?

So next time you look up at the sky at night, remember that you're looking at a whole bunch of amazing celestial bodies.

The Moon is Earth's only natural satellite. This means that it is the only satellite not put into Earth's orbit by humankind.

Let me explain a little about how to think of what a satellite is. Imagine you're throwing a ball to a friend, but instead of throwing it straight to them, you throw it really high up into the air. If you throw it hard enough, it goes up so high that it keeps going and going and never comes back down. That's kind of like what a satellite is. A satellite is a special object that we send up really high into the sky so it can stay up there and do important things for us. Satellites are made by scientists and engineers and can be as big as a school bus or as small as a football. They are usually put into space by rockets, which are kind of like really powerful fireworks that can launch things up into space. Once a satellite is up in space, it can do all sorts of cool things!

Some satellites are like space detectives, helping us learn more about the other planets and galaxies out there. They can snap pictures and send back all sorts of information about the size, shape, and color of the planets!

Other satellites help us keep track of the weather here on Earth. They can tell us when it's going to rain, when it's going to be sunny, and when a big storm is coming. That way, we can be prepared with either our wellies or sunglasses!

And did you know that some satellites even help us use our phones and computers?! That's right, they bounce signals back and forth from space to Earth, so we can make phone calls, send texts, and even watch funny videos online!

So, how cool is that? Satellites are like little helpers that help us explore space, stay safe during storms, and stay connected with our friends and family.

Now let's go back to exploring the Moon together! As you gaze up into the night sky, do you ever see a big, bright, round shape? That's the Moon! It looks different at different times of the month, sometimes it's a full circle, sometimes a crescent shape, and sometimes it's invisible. That's because the Moon is always moving and changing position in relation to the Earth and the Sun.

The Moon is really important for our planet, too. It affects the tides in the oceans, which means the water moves up and down depending on the position of the Moon in the sky. It also lights up our nights, just like the Sun lights up our days.

But did you know that the Moon is actually really far away from us? It's so far that it takes about 27 days for it to orbit around the Earth! Let's explain what an orbit is. An orbit is a path that an object in space takes around another object, like the way the Earth moves around the Sun. It's like a big loop-de-loop that the object travels on, just like how you might ride your bike around the block.

Imagine having a toy car, and you put it on a track that goes around and around in a circle. The toy car will keep going around and around on that track until it runs out of energy or you stop it. In the same way, objects in space keep going around and around in their orbits until something changes, like if they get too close to another object and get pulled off their path. So just like how you might ride your bike in circles, objects in space can also travel on a circular path or orbit around another object.

Now let's continue exploring the Moon together! Even though the Moon looks like it's close enough to touch, it would take us a long time to get there in a rocket ship. But don't worry, we can still learn all about the Moon without going there. Scientists have sent robots and even people to explore the Moon and learn more about it. They've found out that the Moon is covered in craters and mountains and that it has no atmosphere, which means there's no air to breathe!

You see, the Moon is a big round rock that is always up in the sky at night. People have been looking up at the Moon for a very long time and wondering what it would be like to go there. But it wasn't until 1969 that humans had the technology to actually go to the Moon. Neil Armstrong and his astronaut friends went all the way up to the Moon in a special spacecraft called Apollo 11. When they got to the Moon, Neil Armstrong climbed down a ladder and took a step onto the Moon's surface, and that was a really big moment for all of humanity as he was the first person to set foot on the moon - how amazing is that!

While Neil Armstrong was on the Moon, he and his astronaut friends did some really cool experiments and took some amazing pictures. And then they flew all the way back to Earth, where everyone was so excited to hear about their amazing adventure!

Now let's continue our space adventure and keep learning about all the amazing things in our Solar System. Who knows what other cool things we'll discover!

Now you already know kids, it's question time. How excited are you?

1 - What is a celestial body, and how do scientists study them?

(10 seconds to think and answer)

A celestial body is anything in space that isn't on planet Earth.

That means the Moon, the Sun and there are even more celestial bodies out there that we can't see with our eyes, like comets, asteroids, and planets orbiting other stars far, far away.

By looking at the light and radiation they give off, scientists can learn about their size, their temperature, and even what they're made of!

2 - What are some of the things a satellite can do?

(10 seconds to think and answer)

Satellites are like little helpers that help us explore space, predict the weather and stay connected with our friends and family.

3 - Who was the first person to walk on the Moon?

(10 seconds to think and answer)

It was Neil Armstrong

5

The Eclipse

Well, Let's explore together what a Solar Eclipse is!

Have you ever seen a shadow? Of course you have, well during a solar eclipse, the Moon gets in between the Sun and planet Earth, and it casts a shadow on the Earth. It's like the Moon is playing a big game of peek-a-boo with the Sun!

But the funny thing is, the Moon is much smaller than the Sun, so it doesn't always cover the whole Sun. Sometimes it only covers part of it, and that's called a partial eclipse.

Other times, the Moon covers the whole Sun, and that's called a total eclipse.

When there's a total solar eclipse, it gets really dark outside, almost like it's nighttime, and you can see the stars! It's like the Moon is giving the Sun a big hug and making it take a nap.

People all around the world get really excited when there's a solar eclipse, and they might even wear special glasses to look at it safely. Let's explain further why you need to wear special glasses.

First, let's imagine you're going on a fun adventure to a magical land. This land is called the Sun! The Sun is a big, bright ball of fire in the sky that gives us light and warmth.

Now, during a solar eclipse, the Moon comes between the Sun and the Earth. This makes the Sun look like a crescent or a ring in the sky! It's a really cool and rare sight to see. But here's the thing, when you look at the Sun, even during a solar eclipse, it's like staring into a bright lightbulb. And just like how you shouldn't stare at a bright lightbulb, you shouldn't stare at the Sun without special glasses. If you stare at the Sun for too long, it can hurt your eyes and even make you go blind! That's why it's really important to wear special glasses that protect your eyes during a solar eclipse. These glasses are like superhero masks for your eyes! They have special filters that block out harmful rays from the Sun, so you can look at the solar eclipse safely.

So remember, when you're going on your magical adventure to see the solar eclipse, make sure to wear your special glasses to protect your eyes and have a fun and safe time exploring the wonders of the Sun!

You know what time it is kids, now let's begin the question time.

1 - What is a partial eclipse?
   (10 seconds to think and answer)

The Moon is much smaller than the Sun, so it doesn't always cover the whole Sun. Sometimes it only covers part of it, and that's called a partial eclipse.

2 - What is a solar eclipse?

(10 seconds to think and answer)

This is where the moon covers the whole sun.

3 - Why is it important to wear special glasses when looking at a solar eclipse?

(10 seconds to think and answer)

If you stare at the Sun for too long, it can hurt your eyes and even make you go blind.

6

The Solar System

Now we will get to know the planets in order and their names. We will then get to know them better in chapter seven where we will explore each planet in more detail together. Who's excited to get to know the planets?

Then let's head to our next destination together! Our next stop is going to be the planets so buckle up tight and let's continue our adventure!

Did you know that the solar system is like a big family of planets, stars, and other things that all live together in space? It's like a neighborhood in the universe, where everyone has their own house and yard to play in.

As we explored earlier, the biggest thing in the solar system is the Sun.

The planets are like the Sun's children. There are eight planets in total, and they all have their own unique personalities and features. Some are really big, like Jupiter, while others are smaller, like Mercury.

Now we will get to know the planets in order, starting with those closest to the Sun first and ending with the farthest planet from the Sun.

Mercury is the name of the Sun's nearest planet and it's really hot there!

Venus is the second closest planet to the Sun and is often called the "morning star" or "evening star" because it's really bright in the sky. The first celestial body to show in the sky in the evening and the last to vanish at sunrise is Venus. This is why it is called the Morning and the Evening star.

Next, there's Earth! That's where we live! It's the only planet that we know of that has living things like plants, animals, and humans.

The "Red Planet" is known as Mars because it has a reddish color.

After Mars, there's Jupiter, which is the biggest planet in our solar system and has a really big red spot on it that's actually a giant storm!

Saturn has a bunch of pretty rings around it, and Uranus and Neptune are both icy planets that are really far away from the Sun.

Can you believe there is more than one Moon in our amazing solar system? So let's blast off and discover them together!

First, let's talk about the planets that are closest to the Sun. Mercury and Venus don't have any moons, which means they don't have any space rocks orbiting around them. But don't worry. They're still very cool planets to learn about!

Now, let's focus on our home planet, Earth. Did you know that Earth has a special moon that we call "The Moon"?

Yeah that's right, that's what we learned when we met the Moon in chapter three. Do you remember?

It's the only Moon that Earth has, and it's very important to us because it helps regulate the tides in our oceans! Do you know what that means?

Don't worry kids. We got you! Imagine you're at the beach with your family, and you see the water level changing throughout the day. You might wonder why this is happening! Well, the answer is the Moon.

That's right. The Moon is not just a cool thing to look at in the sky; it's also responsible for the tides in our oceans! You might be thinking, "How can a big rock in space affect the water on Earth?" Well, it's all thanks to gravity!

Gravity is like a superpower that some things in space have, including the Moon, it has a big impact on the water in our oceans. When the Moon is in a certain position, it pulls the water towards it which makes the water level go up. This is what we call "high tide"! And when the Moon is in a different position, it doesn't pull the water as much, so the water level goes back down. This is "low tide"!

So, it's like the Moon is playing a game of tug-of-war with the water in our oceans! And it's a really important game because it helps animals like crabs, clams, and fish know when to come out and play and when to go back into hiding. It also helps boats and ships move around more easily.

Isn't that amazing? The Moon has so much power that it affects the entire ocean! And the best part is, you can see it happening right before your eyes at the beach! So next time you're playing in the sand and notice the water level changing, you'll know that it's all thanks to our big, beautiful Moon!

Next, let's zoom over to Mars, which is also a rocky planet like Earth. Mars has not one but two moons! Their names are Phobos and Deimos, and they're both really small compared to our Moon.

But now, get ready for something really exciting! The outer planets in our solar system, called the gas giants and ice giants have lots and lots of moons! Jupiter, which is the biggest planet, has over 80 moons! Can you believe that? Some of its famous moons are named Io, Europa, and Ganymede. These moons are really special because they have volcanoes and oceans and might even have the possibility of life on them!

Saturn, which is famous for its beautiful rings has over 80 moons too! Its biggest Moon is called Titan, and it's even bigger than our Moon. Uranus and Neptune are called ice giants, and they also have lots of moons. Uranus has 27 moons, and Neptune has 14. Scientists are making new discoveries all the time so it wouldn't surprise me if these numbers change in the future.

So, as you can see, the moons in our solar system come in all shapes and sizes, and they're all really cool in their own way! Let's keep on learning and never stop asking questions!

Now let's answer some questions together.

1 - How many planets are there in our solar system?

(10 seconds to think and answer)

There are eight planets in our solar system.

2 - Which planet in our solar system is the biggest?

(10 seconds to think and answer)

Yeah that's right, Jupiter is the biggest, well done!

3 - What is the smallest planet in our solar system?

(10 seconds to think and answer)

You got it right, it's Mercury.

7

The Inner & Outer Planets

Now make sure your seatbelts are buckled up because we're going to learn about the inner and outer planets!

Let's start with the inner planets. Imagine you're standing on a spaceship, and you're flying closer and closer to the Sun. The first planets you'll come across are Mercury, Venus, Earth, and Mars. These four planets are known as the inner planets because they're the closest to the Sun. They're also called "rocky planets" because they're made mostly of rock and metal.

Mercury is the solar system's smallest planet, and it's super hot because it's so close to the Sun. Venus comes in second place as the second planet from the Sun, and it's known for being our solar system's hottest planet because of its thick atmosphere. Earth is our home planet, and it's the ideal distance from the Sun to support life! And Mars is known as the "Red Planet" because of its rusty color, and it's also the planet that scientists are most interested in exploring for signs of life, yes you read that right, signs for life - how awesome is that?

Now, let's talk about the outer planets. These planets are way farther away from the Sun, and they're known as the "gas giants" because they're made mostly of gas and ice. Imagine you're back on that spaceship, and you're flying farther and farther away from the Sun. The first planet you'll come across is Jupiter, which is the biggest planet in our solar system! It's also known for having a giant storm called the "Great Red Spot."

Next up is Saturn, which is famous for its beautiful rings made of ice and rock. After that, you'll come across Uranus, which is known for being tipped on its side, and its rings are almost invisible. And finally, you'll reach Neptune, which is the farthest planet from the Sun and is known for having the strongest winds in the solar system!

So there you have it, young explorers! The inner planets are Mercury, Venus, Earth, and Mars, and they're rocky and close to the Sun. The outer planets are Jupiter, Saturn, Uranus, and Neptune, and they're gas giants and farther away from the Sun. Keep exploring and learning, and who knows what other amazing things you'll discover in our solar system!

Who's excited for question time? Now let's see who will know the answers to these questions.

1 - Which planet is known for being our home planet and is the ideal distance from the Sun to support life?

(10 seconds to think and answer)

You got it right, it's Earth of course.

2 - Why are the four outer planets in our solar system called "gas giants"?

(10 seconds to think and answer)

They're known as the "gas giants" because they're made mostly of gas and ice.

3 - Out of all the planets which is the smallest?

It's Mercury, well done!

8

Exploring The Planets

Hey there kiddos, We meet again! Now we're going to take a closer look at the eight planets and explore more! Are you ready to learn about Mercury, the closest planet to the Sun?

Mercury is super small, even smaller than the Moon we see at night! Now, because Mercury is so close to the Sun, it gets super hot there. So hot that you can fry an egg on it! But wait, don't go packing your bags for a trip to Mercury just yet. There's no air to breathe on Mercury, so it's not a place for us to visit.

But did you know that Mercury has something really cool that no other planet in the solar system has? They're called "scarps," and they're like really tall cliffs that can be over a mile high! Whoa! Imagine climbing a cliff that's taller than the tallest building you've ever seen!

Are you ready to learn about Venus, the "sister planet" to Earth? That's right, Venus is a lot like Earth in size, but it's definitely not a place we'd want to visit. Why you ask? Well, Venus is so hot that it could melt metal! Plus, its air is filled with poisonous gases that we can't breathe. So we definitely wouldn't want to pack our bags for a trip to Venus anytime soon. But here's something really interesting about Venus: it spins backwards! Most planets spin in a clockwise direction, but Venus likes to do things differently and spins counterclockwise. It's like Venus is a rebel planet! So even though we can't visit Venus because it's way too hot and dangerous for us, we can still admire its uniqueness from afar. Who knows, maybe one day we'll have a spaceship that can take us there safely to learn more about this rebel planet!

Now let's head to Earth! Earth is the planet we call home, and it's a very special planet indeed. Earth is the only planet that has living things on it, like plants, animals and people. It's a beautiful planet with lots of different landscapes, like mountains, oceans, and forests. And, it's just the right distance from the Sun, which means it's not too hot or too cold for us to live on. But we need to take care of our planet because it's the only one we have. That's why we recycle, use energy wisely, and try to keep our air and water clean. We want to make sure that Earth stays healthy and beautiful for generations to come!

Have you ever heard of recycling? It's a super cool way to help take care of our planet and make sure we keep it healthy for a long time to come.

Recycling is when we take things that we've used, like plastic bottles or cardboard boxes, and turn them into something new. We do this by putting these items in special recycling bins instead of throwing them in the trash. These recycling bins are like magic boxes that can turn old things into new things. For example, a plastic bottle could be turned into a new t-shirt or even a backpack! And when we recycle, we're not only helping the planet; we're also saving resources like water and energy.

So, next time you're finished with your juice box or your cereal box, think about putting it in the recycling bin instead of the trash. You'll be doing your part to help take care of our planet and keep it healthy and beautiful for a long time to come!

So let's pause a moment to appreciate our amazing planet Earth and all the living things that call it home. We can all do our part to take care of our planet by keeping it clean and healthy. After all, it's the only home we have!

I want you to close your eyes now for a little bit and think about all the amazing things we have on planet earth and how grateful we should be.

Feeling good, good! Are you now ready to learn about Mars, also known as the "Red Planet"? That's right, when we look up at the night sky, we can see Mars shining bright and red like a cherry on top of an ice cream sundae!

Mars is a cold and dry planet, but it's super special because it's the planet that scientists think is most like Earth. Wow! Can you imagine that? Maybe one day we can even live on Mars and have space houses just like we have houses here on Earth. In fact, NASA (the space exploration agency) has sent special robots called "rovers" to Mars to study it up close and personal. These rovers are like special space cars that can drive around and take pictures and collect rocks to bring back to Earth. Scientists want to learn more about Mars and are working hard and making an effort to figure out how to make Mars a habitat for humans. Do you know what habitats are? Let's find out together! Have you ever wondered where animals live and how they survive? Well, every living thing has a place where it lives, called a habitat. A habitat is like a home for animals and plants. Just like you have a house or apartment where you live, animals and plants have habitats too!

A habitat can be anywhere in the world, like a forest, a desert, a river, or even your own backyard! Each habitat has unique features like temperature, water, and food sources that help different animals and plants survive and thrive.

Are you curious to know where humans live?

Well, we live in something called a "habitat" which is just a fancy word for our homes!

A habitat is a special place that we create for ourselves to keep us safe, cozy, and happy. Just like animals have their homes in the jungle, in the ocean, or even in trees, we humans have our habitats too! But unlike animals, we can create and design our habitats in many different ways.

Some humans live in big buildings called apartments or flats, where many families can live together in the same place. Other humans live in houses with gardens where they can play and enjoy the sunshine. Some humans even live in houses on wheels that they can move from one place to another! Isn't that cool?

But no matter where we live, we all want our habitats to be comfortable and have things that we need, like a bed to sleep in, a kitchen to cook our food, and a bathroom to clean ourselves. We also like to decorate our habitats with things that make us happy, like pictures, toys, or plants.

So, little friends, human habitats are just the special places we call home! And whether we live in a big city or a small village, our habitats are where we feel safe and loved by our families and friends. Each habitat is unique and has its own plants and animals that are adapted to live there. It's important to take care of habitats so that all creatures can have a safe and healthy home. We can all help in some way to protect habitats by planting trees and recycling.

So, did you have fun exploring different habitats? Remember, habitats are like homes for animals and plants, and each one is unique and special!

Mars is a planet that is similar to Earth in many ways. It has a rocky surface, mountains, valleys, and even polar ice caps, just like Earth! However, there are some big differences between the two planets. Mars is very, very cold and has a very little atmosphere, which means there is no air to breathe. But don't worry; scientists are working hard to find ways to live on Mars!

One of the things scientists are doing is trying to create a way to make oxygen on Mars. Oxygen is the gas we breathe, and we need it to live. So, scientists are working on ways to make oxygen on Mars by using the resources that are available there, like the carbon dioxide in the atmosphere. Carbon dioxide is a gas that we can't see or smell, but it's all around us, especially in the air we breathe. We create carbon dioxide when we breathe out, and many things like cars and factories also produce it when they burn fuel. While some carbon dioxide is necessary for plants and animals to survive, too much of it can be harmful to the environment, causing things like climate change. Do you know what climate change is? Let's find out together!

The Earth has a big blanket of air called the atmosphere that keeps it warm. When we burn things like coal, oil, and gas, it makes the Earth too warm and changes the climate, making it hotter or colder than it should be. This can be bad for plants and animals, and even for people, so we need to be careful about how much we burn and try to use clean energy like the Sun and wind instead.

Climate change is a big problem that is affecting the Earth in many ways. When the Earth gets too warm, it can cause things like droughts, floods, and storms that can hurt plants, animals, and people. It can also cause the polar ice caps to melt, which can make the sea level rise and flood some of the lands.

We can help stop climate change by doing things like using less electricity, walking or biking instead of driving in a car, and recycling things we use. We can also try to plant more trees, which take in carbon dioxide from the atmosphere and release oxygen in exchange. So next time you take a deep breath of air, give credit to a tree or hug a tree in thanks for what it gives us. By working together, we can make a big difference in helping to protect our planet and keep it healthy for everyone to enjoy. Now let's continue exploring the possibility of life on Mars!

Another thing scientists are doing is figuring out how to grow food on Mars. Food is very important for us to survive, and it's not easy to grow things on a planet that doesn't have much water or nutrients in the soil. But, scientists are working on ways to grow plants in special containers that have all the things plants need to grow.

In the future, we might even be able to make a human colony on Mars! A colony is like a little village where people live and work together. But first, we need to figure out how to get there and how to live there safely.

Scientists are working on building rockets and spaceships that can take us to Mars, and they're also figuring out how to protect us from things like radiation and extreme temperatures.

So, in summary, scientists are working hard to figure out how to make oxygen, grow food, and build a human colony on Mars. We have a lot to learn, but it's a very exciting time for space exploration! Who knows, maybe one day you could be one of the scientists working on Mars. Imagine that!

So even though Mars might be far away and a little bit cold, it's still an exciting planet to learn about and maybe one day we'll get to visit it ourselves!

Are you ready to learn about Jupiter, the king of the planets.

Jupiter is so large that all the other planets could fit inside of it. And still have room left over!

But that's not all! Jupiter is also a really windy planet. In fact, the winds on Jupiter can blow up to 400 miles per hour! That's like a super-fast rocket ship going zoom!

And did you know that Jupiter has over 80 moons? There's no other planet in the solar system that has more than that! Also, some of these moons are even bigger than the planet Mercury - can you believe it?

Now, get ready to learn about Saturn, the planet with the most bling in the solar system! Saturn is famous for its beautiful, sparkly rings that make it look like it's wearing a fancy belt. These rings are made up of tiny particles of ice and rock, and they're so big that you can see them from Earth with a special telescope.

Saturn is also a really big planet - almost as big as Jupiter, the king of the planets. That's a whole lot of planets to explore! It has been discovered that Saturn too has over 80 moons. That's a lot of moons as you can imagine and every single one is unique and special in its own way.

Now, get ready to learn about Uranus, the planet that likes to do things a little differently! Uranus is a planet that likes to have fun and stand out from the crowd. Instead of spinning up and down like most planets, Uranus spins on its side like it's doing a cartwheel around the Sun!

Uranus is also known for being super cold - even colder than Neptune, it's neighbor. But despite the chilly temperatures, Uranus has a really pretty blue color that makes it stand out in the sky. That's because of all the special gases in its atmosphere that reflect the Sun's light and make it look like a giant blue marble.

Now, let's take a fun journey through our solar system and learn about Neptune, the farthest planet from the Sun!

It's super cold! Brrrr! It's so chilly on Neptune because it's the farthest planet from the Sun and because it's super cold, you would need a big jacket, a hat, and gloves to stay warm. But that's not all! Neptune is also one of the windiest planets in our solar system. The winds there are so strong that they could lift you up like a kite and carry you away! And guess what? Neptune has a big, dark spot on its surface that's like a giant hurricane! It's like a never-ending dance party with crazy winds and wild weather.

Now, let's talk about Neptune's friends. Neptune has 14 moons, and one of them is special because it's bigger than all the others. Its name is Triton, and it's like the coolest Moon in town!

Triton is an icy moon, just like a popsicle, with a super thin atmosphere that's like a soft blanket. It's so cold on Triton that you could freeze an ice cream cone in seconds! The temperature can drop as low as -235 degrees Celsius! That's colder than your freezer at home!

Even though Triton is one of Neptune's 14 moons, it's actually one of the biggest moons in our whole solar system. Scientists think that Triton may have been a dwarf planet that was captured by Neptune's gravity. It's like Neptune saw Triton and said, "Hey! You're so cool; I want you to be my friend!"

Triton's surface is like a big candy store, covered in frozen nitrogen, and it has geysers that shoot nitrogen gas and dust particles into space. It's like a big science experiment happening right before our eyes!

Even though Triton is a super cold and crazy moon, it's taught us so much about our solar system. So, next time you look up at the night sky, remember Neptune and its cool friend Triton and all the secrets they hold!

We've reached the end of this chapter so you know what this means. It's question time again.

1 - Why is Mercury not a place for humans to visit?

(10 seconds to think and answer)

Yeah that's right, There's no air to breathe on Mercury, so it's not a place for us to visit.

2 - What is recycling?
   (10 seconds to think and answer)

Recycling is when we take things that we've used, like plastic bottles or cardboard boxes, and turn them into something new. We do this by putting these items in special recycling bins instead of throwing them in the trash.

3 - This is a tough one now, can you remember how many moons Neptune has?

It's 14 - isn't that amazing!

9

Pluto & The Kuiper Belt

Did you know that Pluto used to be considered the ninth planet in the incredible solar system, but now it's not?

Let's get to know Pluto better!

Pluto is a tiny, cold and far away planet that is part of our solar system. It's really far away from the Sun, so it's always very cold there. It's so cold that everything on Pluto is made of ice, like ice mountains and ice volcanoes!

Pluto is also very small, so small that it's smaller than our own Moon! That means that it doesn't have a lot of gravity, so if you were standing on Pluto, you could jump really high! That would be pretty cool right?

Scientists learned more about Pluto and found out that it's actually part of a group of small, icy objects called the Kuiper

Belt. But even though it's not considered a planet anymore, Pluto is still very interesting and important to study!

One really fun fact about Pluto is that it has a moon named Charon. Charon is almost as big as Pluto itself! And because Pluto and Charon are so close in size, they actually orbit around each other instead of just one orbiting around the other, as the Earth and Moon do.

Another cool thing about Pluto is that it has a really long year. A year on Pluto is almost 250 Earth years! That means that if you lived on Pluto, you would only have one birthday every 250 years! I bet you're glad you live on Earth.

So, that's a little bit about Pluto, the tiny, cold, and fascinating planet (or former planet) at the edge of our solar system!

Scientists used to consider Pluto a planet, but now it's known as a "dwarf planet." This happened because scientists discovered that Pluto wasn't alone in the outer parts of our solar system and that there are other similar objects like Pluto that also orbit the Sun in the Kuiper Belt.

The International Astronomical Union (IAU) in 2006, decided to reclassify Pluto as a "dwarf planet." This means that it is a special type of planet that is smaller than the other eight planets in our solar system and that together with other objects in the Kuiper Belt, it orbits in a shared space.

Even though it's no longer called a planet, Pluto is still very important to study! Scientists have sent spacecraft to study Pluto and its moons, and they've learned a lot about this distant and fascinating world. In fact, some scientists think that there could be many more dwarf planets like Pluto in our solar system that we haven't discovered yet!

So, even though it's not considered a full-fledged planet anymore, Pluto is still a very interesting and important object in our solar system.

Are you ready to explore the Kuiper Belt?

Let's explore together!

The Kuiper Belt is a special place in our solar system that's located beyond the orbit of Neptune, which is the eighth planet from the Sun. It's a big, donut-shaped region filled with lots of icy objects like comets, asteroids, and dwarf planets, including Pluto!

Think of it like a giant space neighborhood that's home to a lot of smaller objects. Some of these objects are really tiny, like pebbles or rocks, and others are much bigger, like Pluto!

Scientists believe that the Kuiper Belt is very important because it contains leftover materials from the formation of our solar system. By studying these icy objects, scientists can learn more about how our solar system was created and how it has changed over time.

So, the Kuiper Belt is a special place in our solar system that's home to lots of icy objects, including dwarf planets like Pluto. It's an important area for scientists to study because it can teach us more about how our solar system was formed.

Okay kids, it's question time again. Who's been paying attention? Let's see who can answer the following questions:

1 - Why is the Kuiper Belt important for scientists to study?

(10 seconds to think and answer)

It's an important area for scientists to study because it can teach us more about how our solar system was formed.

2 - How many Earth years are in one year on Pluto?

(10 seconds to think and answer)

One year on Pluto is almost 250 Earth years.

3 - True or false question here - was Pluto considered to be the 9th Planet in our solar system?

That is true - well done!

10

The Shooting Star

Hey there, my little space adventurers! This is a super short chapter, but it's all about something really cool - shooting stars!

Now, you might be thinking, "What's the big deal? There's nothing much to talk about here." But trust me, shooting stars are something special.

Picture this: you're outside on a dark, starry night, gazing up at the sky. All of a sudden, you see a bright light shoot across the sky like a sparkler. That's a shooting star! And even though it only lasts for a few seconds, it's like a magical moment that you'll always remember. But guess what? Shooting stars aren't really stars at all! They're actually little pieces of rock and dust from space that are zooming through our atmosphere. And when they do, they get super hot and start to glow like a cosmic campfire.

It's like when you rub your hands together really fast, and they get warm. But shooting stars are going so fast that they get even hotter than your hands ever could! That's what makes them look like a shiny star in the sky.

We're not done yet kids. You'll answer the last question first then after that we'll get to the end of our journey when we reach the conclusion together.

What are shooting stars and how are they formed?

(10 seconds to think and answer)

Shooting stars aren't really stars at all. They're actually little pieces of rock and dust from space that are zooming through our atmosphere. And when they do, they get super hot and start to glow like a cosmic campfire.

11

Conclusion

Wow, we've reached the end of our incredible journey through the solar system! From the blistering heat of the Sun to the distant reaches of Pluto, we've learned so much about the fascinating world beyond our own. I hope this book has sparked your curiosity and inspired you to continue exploring the mysteries of space.

Remember, the universe is vast and full of wonders waiting to be discovered. So keep looking up at the stars and dreaming big!

Thank you for joining us on this awesome adventure, and we hope you'll join us again for more exciting journeys.

12

References

Book Cover Image: Freepik.com - This cover has been designed using assets from Freepik.com

Chapter 1:

Smith, M. (2022, May 23). Comets vs asteroids: How do these rocky objects compare? Space.com. https://www.space.com/comets-vs-asteroids

What Is a Nebula? | NASA Space Place – NASA Science for Kids. (n.d.). https://spaceplace.nasa.gov/nebula/en/#:~:text=gas%20in%20space.-,Some%20nebulae%20(more%20than%20one%20nebula)%20come%20from%20the%20gas,are%20called%20%22star%20nurseries.%22Chapter

Chapter2:

REFERENCES

Zuckerman, C. (2021, May 3). Everything you wanted to know about stars. Science. https://www.nationalgeographic.com/science/article/stars#:~:text=Stars%20are%20huge%20celestial%20bodies,all%20light%2Dyears%20from%20Earth

Chapter 3:

How Often is "Once in a Blue Moon"? (n.d.). Wonderopolis. https://www.wonderopolis.org/wonder/How-Often-is-%E2%80%9COnce-in-a-Blue-Moon%E2%80%9D

Cain, F. (2015, November 23). Did we need the moon for life? https://phys.org/news/2015-11-moon-life.html

Waxman, O. B. (2019, July 15). Lots of People Have Theories About Neil Armstrong's "One Small Step for Man" Quote. Here's What We Really Know. Time. https://time.com/5621999/neil-armstrong-quote/

Chapter 5:

The solar system. (n.d.). Khan Academy. https://www.khanacademy.org/science/middle-school-earth-and-space-science/x87d03b443efbea0a:earth-in-space/x87d03b443efbea0a:the-solar-system/v/the-solar-system-ms

Venus. (n.d.). NASA Solar System Exploration. https://solarsystem.nasa.gov/planets/venus/overview/

B. (2022, July 4). Why is Venus called the Morning and the Evening Star-. https://byjus.com/question-answer/why-is-venus-called-the-morning-and-the-evening-star/

Overview | Mars – NASA Solar System Exploration. https://solarsystem.nasa.gov/planets/mars/overview/

Bài 4. Máy tính và ph`n m`m máy tính - Tin học 6 - cao minh phát - Thư https://baigiang.violet.vn/present/may-tinh-va-phan-mem-may-tinh-13554382.html

Chapter 6:

Mercury. (n.d.). NASA Solar System Exploration. https://solarsystem.nasa.gov/planets/mercury/overview/

Eight ingredients for life in space. (n.d.). Natural History Museum. https://www.nhm.ac.uk/discover/eight-ingredients-life-in-space.html

Chapter 7:

Astronomy for Kids: The Planet Earth. (n.d.). https://www.ducksters.com/science/earth.php

Lifebabble - Help me out - going green. (2017, March 2). https://www.bbc.co.uk/cbbc/findoutmore/help-me-out-going-green

Jupiter Facts. (n.d.). https://kidzone.ws/planets/jupiter.

htm

Jupiter's moon count jumps to 92, most in solar system. (2023, February 7). https://wsvn.com/news/local/florida/jupiters-moon-count-jumps-to-92-most-in-solar-system/

Chapter 8:

Atkinson, N. (2018, May 12). Order Of the Planets From The Sun - Universe Today. Universe Today. https://www.universetoday.com/72305/order-of-the-planets-from-the-sun/

S. (2023, February 18). Hello Pluto. . . Science and Maths With SciTeach. https://sciteach.substack.com/p/hello-pluto

Remple, A. (2023, January 13). When Was Each Planet Discovered? WorldAtlas. https://www.worldatlas.com/space/when-was-each-planet-discovered.html

Hess, C. (2022, November 21). Engaging Kids Solar System Activities and Lesson Plans. Hess Un-Academy. https://hessunacademy.com/kids-solar-system/

Once considered the ninth planet in the solar system, Pluto was discov. (n.d.). GMAT Club Forum. https://gmatclub.com/forum/a-brand-new-question-from-e-gmat-172543.html

planet. (n.d.-b). https://www.nationalgeographic.org/encyclopedia/planet/

Manufactured by Amazon.ca
Bolton, ON